Sur...art

Animals and plants grow and chan... as babies. They grow into children...

Read each set of sentences below. Number them in order.

A.

____ The tadpole loses its tail.

____ The tadpole grows back legs and front legs.

____ A tadpole hatches from an egg.

____ A frog lays tiny eggs in a pond.

____ The young frog begins to live its life on land.

B.

____ The young grasshoppers eat plants.

____ The eggs remain warm underground, and the babies inside the eggs grow.

____ Baby grasshoppers climb out of the hole in the ground.

____ The young grasshoppers grow into adults.

____ The eggs hatch.

____ A grasshopper lays eggs in a hole in the ground.

Next!

Read the paragraphs below about life cycles of animals. Fill in the circle that tells what happens next.

1. A female mosquito lays 100–300 eggs. The eggs hatch into very active larvae. The larvae look like worms. The larvae change into pupae. In two to four days, the shell of the pupa splits, and . . .

   ○ a mosquito flies out.

   ○ a larva crawls inside.

2. Two birds build a nest. The female lays eggs. She sits on the eggs to keep them warm. The male brings her food. The chicks hatch. The mother and father feed the chicks. The chicks grow bigger and bigger. They grow feathers.

   ○ The babies stay in the nest with their parents.

   ○ The babies learn to fly, and then they leave home.

3. A female dog has puppies. The puppies drink their mother's milk. Write three complete sentences that tell what will happen next.

   _____

   _____

   _____

Staying Alive

Each day, most animals must look for other animals they can eat. They must also hide from animals that would eat them. Camouflage helps animals hide from predators. It can also help them hide from their prey.

Write a sentence from the box to finish each cause or effect sentence about camouflage.

Cause

1. _____
   _____

2. Because the hawk moth caterpillar has a horn that looks sharp,

3. Because a green frog sat on a green lily pad,

4. Because the ladybug was red and black, the colors of some poisonous insects,

Effect

1. The hawk couldn't see the rabbit.

2. _____
   _____
   _____

3. _____
   _____

4. _____
   _____
   _____

The white rabbit sat still on the snow.

a predator would leave it alone.

the bird did not eat it.

the frog caught an insect that hadn't seen the frog.

In Self-defense

Some animals have built-in ways to defend themselves from predators. They can bite, sting, make bad smells, hide in their own shell, or taste nasty. Others find clever ways to protect themselves and their young.

Use the code to complete the sentences to learn about animal defenses.

A = 1	D = 4	G = 7	J = 10	M = 13	P = 16	S = 19	V = 22	Y = 25
B = 2	E = 5	H = 8	K = 11	N = 14	Q = 17	T = 20	W = 23	Z = 26
C = 3	F = 6	I = 9	L = 12	O = 15	R = 18	U = 21	X = 24	

1. A __ __ __ stings its enemies and its prey.
      2 5 5

2. A __ __ __ __ __ __ __ __ __ __ shakes the rattles in its tail to
     18 1 20 20 12 5 19 14 1 11 5
   warn away predators.

3. A __ __ __ __ __ __ __ __ __ __ pulls into the shell it carries on
     8 5 18 13 9 20 3 18 1 2
   its body.

4. A __ __ __ __ __ __ fish swallows air or water to make itself look
     16 21 6 6 5 18
   larger than it is. It floats upside down until it feels safe.

5. A __ __ __ __ __ __ can pull into the shell it carries on its body.
     20 21 18 20 12 5

6. A __ __ __ __ __ can spray a bad smell at enemies.
     19 11 21 14 11

7. A __ __ __ __ __ __ __ __ __ __ __ can spurt blood from the
     8 15 18 14 5 4 12 9 26 1 18 4
   corners of its eyes to scare away predators.

© Frank Schaffer Publications, Inc.

Extreme Conditions

Animals live in all kinds of extreme conditions.

Complete the puzzle to learn how some animals must find ways to live in some extreme conditions. Use the words in the box to help you.

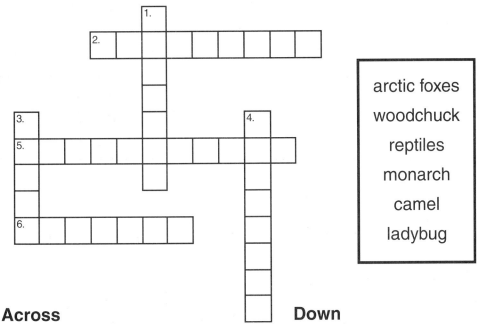

arctic foxes
woodchuck
reptiles
monarch
camel
ladybug

Across

2. This small animal eats all summer and gets fat. In cold places, it sleeps all winter. Its heart slows, and its body uses the stored fat.

5. These animals have small feet and ears to keep them from losing heat in the icy world in which they live.

6. This small insect sleeps in a protected place during cold winter weather to avoid freezing.

Down

1. The caterpillars of this butterfly eat milkweed, a poisonous plant. This makes them taste bad to birds and other animals.

3. This animal lives in hot, dry parts of the world. It can go for days without water. It stores flesh and fat in its hump.

4. In parts of the world where it gets cold in winter, these cold-blooded animals sleep all winter. Snakes belong to this group.

Your Mother's Eyes

Every child has two parents. Each parent has genes that give him or her a certain color of hair, skin, and eyes. These genes also determine each child's height, body shape, and other details. A child carries genes from both parents. Combined, the genes make up a new, unique person.

Look at the picture of the family and answer the questions.

1. Which parent did this child's eye shape come from? _____
2. Which parent did she get her hair color gene from? _____
3. From which parent did she get the shape of her nose? _____
4. From which parent did she get her freckles? _____
5. From which parent did she get her mouth shape? _____
6. The parents' genes determined that the child would be what kind of animal? _____
7. List three traits common to all humans. _____
   _____
   _____

It's in the Genes

Children look like their parents and grandparents. Other animals look like their parents, too.

Look at the picture of the two dogs and their puppy. Because these animals are dogs, their baby will be a dog. The parents give the baby the genes that make it a dog. Answer the questions about the puppy's genes.

1. From which parent did the puppy get its spots? _____
2. From which parent did the puppy get its nose size and shape? _____
3. List one other trait the puppy got from its mother. _____
4. List one other trait the puppy got from its father. _____
5. Tell two ways the father is different from the mother. _____
_____
6. Tell one way the mother is different from the father._____
_____
7. Tell one way the parents are alike. _____
_____
8. List three traits common to all dogs._____
_____
_____

Biography of a Giant Sequoia

A sequoia is an evergreen tree that can grow to huge heights! It is not the tallest tree on Earth, but it weighs many tons. These trees grow only in the western part of the Sierra Nevada. Insects and fires have little effect on this hardy tree.

Number the sentences below in order from the beginning of life for a sequoia to the time when it is full grown.

_____ Rain and sun fall on the ground, and the young tree pokes through the soil.

_____ A tiny seed falls on the ground.

_____ The tree grows for 3,000 years or more and weighs more than a thousand tons.

_____ The young tree develops brown bark and green leaves.

Name three traits all sequoia trees have.

1. _____

2. _____

3. _____

Name That Habitat

Habitats are areas in which plants and animals live. Each habitat has a particular climate, certain kinds of weather, and other features. Very large habitats are called biomes.

Below are five habitats. Use the words in the box to help you write the name of each kind of habitat on the line beside it.

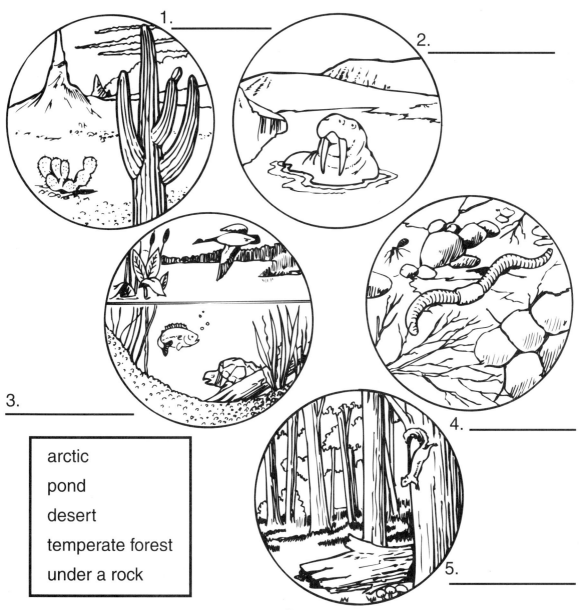

arctic
pond
desert
temperate forest
under a rock

Your Own Backyard

It may surprise you to discover that places you visit every day are habitats for many plants and animals. Think about your backyard, schoolyard, or a nearby park. Choose one of these habitats. Follow the directions below.

1. Write the name of an animal from this habitat on the top line in each box in the first row.
2. Draw and color a picture of each of these four animals.
3. On the bottom line, label the correct group of each animal (bird, mammal, reptile, spider, or insect).
4. Write the name of a plant on the line in each box in the second row.
5. Draw and color a picture of each of the plants you listed.

A N I M A L S

P L A N T S

What kind of habitat is it? _____ Describe it. _____

Circle of Life

A habitat is a place where animals and plants live. It is a kind of community. The animals in a habitat eat the plants or other animals that live there. If some animals or plants disappear from a habitat, other animals won't have enough to eat. A pond is a habitat. Match each statement about a pond animal to its picture.

1. The nymph of this insect lives in water and eats fish and insects.

2. This fish eats many other animals in the pond. It has whiskers.

3. At this stage of a frog's life, it lives in water. A frog is an amphibian.

4. This reptile has strong jaws and bites hard. It eats fish.

5. This reptile hunts for and eats frogs and fish.

6. Most of this kind of mollusk eat plants and dead animals.

The Food Web

In a food web, plants and animals all play a part in a cycle of eating and being eaten. It begins with the sun, which sends light energy to Earth. Plants use the light energy to make their own food. Some animals eat plants for energy. These animals are called herbivores. Some animals get their energy by eating herbivores. These animals are called carnivores. Animals that eat both plants and animals are called omnivores.

In your own words, describe what is happening in each step of this food web. Use these three words: **herbivore**, **carnivore**, **omnivore**.

Who's Eating What?

An ecosystem is another word for habitat. All parts of an ecosystem depend on each other. In each ecosystem, there are living parts—producers, consumers, and decomposers. The producers are green plants. The consumers are herbivores, carnivores, and omnivores. Decomposers and dead plants and animals form another part of the system.

Use the code to fill in the blanks. For the second part of each question, tell whether the animal is a **carnivore**, an **herbivore**, or an **omnivore**.

A = 1	D = 4	G = 7	J = 10	M = 13	P = 16	S = 19	V = 22	Y = 25
B = 2	E = 5	H = 8	K = 11	N = 14	Q = 17	T = 20	W = 23	Z = 26
C = 3	F = 6	I = 9	L = 12	O = 15	R = 18	U = 21	X = 24	

1. A __ __ __ __ __ eats mice, rats, insects, and seeds. It is a(n)
   19 11 21 14 11
   _____.

2. A __ __ __ __ __ eats plants. It is a(n) _____.
   19 14 1 9 12

3. A __ __ __ __ eats seals. It is a(n) _____.
   2 5 1 18

4. A __ __ __ __ __ eats only eucalyptus leaves. It is a(n)
   11 15 1 12 1
   _____.

5. A __ __ __ __ __ __ __ __ __ __ __ eats berries and insects. It is
   13 15 3 11 9 14 7 2 9 18 4
   a(n) _____.

© Frank Schaffer Publications, Inc. FS111161 Science

Desert Life

In the picture below are many animals that live in a North American desert. Follow the directions to see how much each animal depends on the other animals and plants in its habitat for survival. Plants and animals in any habitat need each other to survive.

1. List the animals that breathe in oxygen that plants give off. _____
_____

2. Circle blue everything that needs water to live.

3. Draw a red line under each animal that eats plants or animals to live.

4. Put a yellow X on everything in the picture that needs sunlight to live.

5. Write a sentence explaining how all animals need plants. _____
_____

Super Skills

Many animals are born with skills that they use to fix their environment in ways to help them survive. For example, humans build homes to protect themselves from weather conditions and other dangers. Match each description to the correct picture to learn about some animal skills.

1. Beavers build dams with sticks and mud. The dam holds in water, forming a pool in which the beavers can catch fish. They sleep during the day. At night, they fish and cut down trees for their dams.

2. A hermit crab backs into a shell which protects its body. As the hermit crab grows, it becomes too large for its shell. Then the crab looks for a bigger shell where it goes to live.

3. Each year, the male bowerbird builds a den of sticks decorated with berries, flowers, shiny pieces of metal, and other bright objects to attract a female.

4. People who live in dry parts of the world bring in water from other areas to grow plants. They may use sprinklers, tubes, or furrows to deliver the water. Without irrigation, they cannot grow many kinds of plants.

Change Can Be Bad

In a habitat, each kind of animal and plant depends on the others for life. If something happens in the habitat, many changes occur. To find out what happens if there is too little rain, read each sentence below and decode the names of the animals.

A = 1	D = 4	G = 7	J = 10	M = 13	P = 16	S = 19	V = 22	Y = 25
B = 2	E = 5	H = 8	K = 11	N = 14	Q = 17	T = 20	W = 23	Z = 26
C = 3	F = 6	I = 9	L = 12	O = 15	R = 18	U = 21	X = 24	

1. If there is too little rain . . .

    a. oak trees will die, and there will be no more acorns for the __ __ __ __ __ __ __ __ __.
       19 17 21 9 18 18 5 12 19

    b. plants will die, leaving no food for __ __ __ __
       13 9 3 5
       and __ __ __ __ __ __ __.
            18 1 2 2 9 20 19

    c. the soil will dry up, __ __ __ __ __ will die,
       23 15 18 13 19
       leaving no food for __ __ __ __ __ __.
                           18 15 2 9 14 19

    d. bushes with berries will die, leaving no food for __ __ __ __.
       4 5 5 18

2. Without small animals to eat, __ __ __ __ __, __ __ __ __ __, and
   6 15 24 5 19   8 1 23 11 19
   __ __ __ __ will die.
   15 23 12 19

3. In complete sentences, tell what would happen to this forest full of plants and animals if one big change happened.

   _____

   _____

It's in the Air

In the desert scene below, the air is not clean. Answer each question.

1. Plants need clean water and air to grow. Tortoises eat plants. What happens if the polluted air and water kill the plants this animal eats?
   _____

2. Kangaroo mice live in the roots of plants. They come out at night to eat insects and seeds. What happens to this animal if the plants in the desert die from polluted air? _____

3. Snakes eat small animals. What happens to the snakes if the mice and all other small animals die? _____

4. A roadrunner is a bird that eats snakes and lizards. What happens to this bird if snakes die from lack of food? _____

5. Write three sentences telling why we need to keep the air and water in every habitat clean. _____
   _____
   _____
   _____

Fact or Opinion: Habitats

A fact is something that has happened or is real. *(You breathe oxygen.)* An opinion is something someone believes or thinks. *(You like squirrels.)*

Underline each sentence that is a fact about a habitat. Circle each opinion.

1. Plants need clean air to flourish.

   Healthy plants grow nicely in polluted air.

2. All animals like to eat other animals.

   Some animals eat plants, and others eat animals.

3. A large habitat is called a biome.

   A wet biome is the best place for animals to live in.

4. If plants in a habitat die, eventually, both plant- and animal-eating animals will die.

   If plants in a habitat die, only plant-eating animals will die. The other animals will live.

5. If the air is polluted, only plants will get sick.

   Polluted air affects both plants and animals.

6. People are not affected by polluted water.

   Polluted water is unhealthful for all animals and plants.

What's There?

Choose a habitat. Find out what kinds of plants and animals live in this habitat. Be sure to find out about insects, birds, plants, mammals, reptiles, amphibians, fish, mollusks, and spiders (arachnids).

Color in a square on the bar graph to show how many of each kind of life you found in the habitat you chose.

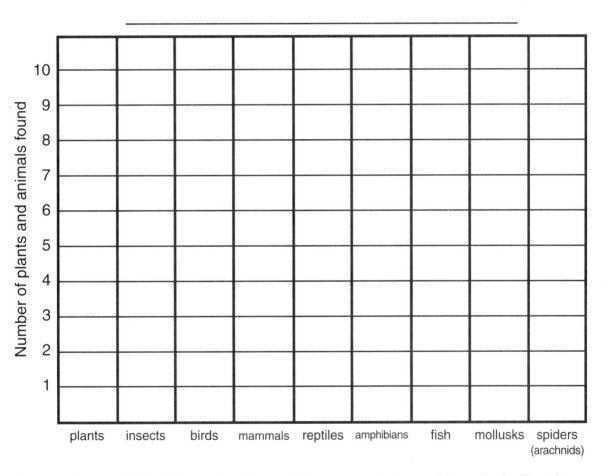

Remember, all birds have feathers. All mammals have fur or hair. Reptiles are cold-blooded animals and include snakes, lizards, and turtles. Mollusks include snails, clams, oysters, and squid. Fish have gills for breathing air under water. Amphibians live in water as babies but live on land as adults. These include salamanders, frogs, and toads.

© Frank Schaffer Publications, Inc.

A Hot Topic

Earth changes continually. Volcanoes bring about one type of change on Earth. A volcano begins as **magma**, melted rock inside Earth. When magma mixes with gas, sometimes it erupts through an opening called a **vent**. Air and water cool the **lava**, which hardens. Each time lava flows from an opening, it hardens again. This hardened lava piles up into a shape called a **cone**. Sometimes openings called **fissures** occur on the sides of volcanoes. Lava flows through fissures, too. Some volcanoes are low and broad, and others are tall.

Use the words in bold above to label the parts of the volcano below.

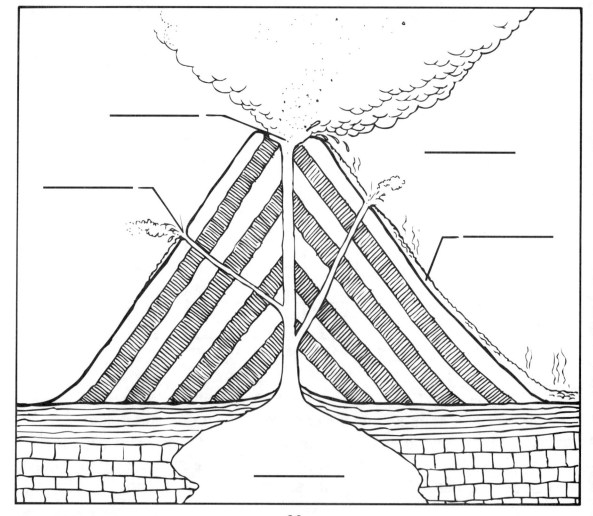

Shake, Rattle, and Roll

Earth's surface is made up of huge plates we call tectonic plates. These plates move a tiny amount each day. At the point where two plates meet, a fissure may occur and lava will flow, perhaps beginning a volcano. If two plates collide, the surface of Earth will move more quickly than normal. This movement is an earthquake. Earthquakes can cause a lot of destruction.

Read the paragraphs below about earthquakes. Fill in the circle in front of the sentence that tells what will happen next.

1. It has rained for several days. Suddenly, an earthquake strikes. The soil in the cliff above the road is heavy with water that has soaked into the ground.

    ○ The soil in the cliff dries suddenly in the heat.

    ○ The heavy soil in the cliff slides down on the road below.

2. An earthquake happens deep in the ocean. A large wave called a tsunami begins to build and move quickly toward land.

    ○ When the tsunami hits land, it washes away houses, other buildings, cars, and trees.

    ○ Tsunamis never hit land.

3. An elevated highway begins to shake in an earthquake. Cars and trucks come to a stop.

    ○ People run out on the road to tell each other about the shaking.

    ○ Part of the highway has fallen down.

Weathering and Erosion

Weathering and erosion change our Earth. Wind, water, and ice weather the land when they take away a top layer of soil. When this happens, erosion begins. Weathering can occur when water runs over land. If the soil can absorb no more water, the excess water runs off the top and takes a layer of soil with it. These types of weathering make it hard for plants to grow.

Wind can also blow away topsoil that is not held in place by plants. Over time, constant erosion lowers the land. Then, if a glacier moves slowly forward, the rocks in the ice grind any solid rock the glacier moves across. Erosion caused by glaciers can carve deep valleys.

1. Write one sentence that tells how glaciers and wind are the same. _____ _____

2. Write a sentence that tells how water and wind can cause weathering in similar ways. _____ _____

3. Write a sentence that tells how water can weather the land. _____ _____ _____

4. Write a sentence that tells what happens when a glacier moves over land. _____ _____

Humans Can Cause Weathering

Humans can cause weathering when they change the land. This, in turn, affects plants and animals. Learn about some causes and effects of weathering by writing a sentence from the box to finish each cause or effect.

> Heavy jungle rains wash away the topsoil.
>
> Wind hollows out the hill on either side of the road.
>
> Someone scrapes all grasses and plants from the land.
>
> The land becomes a desert.
>
> turned the land into desert.

Cause

1. Someone cuts down all the trees in a jungle.

2. _____

3. Cattle graze on land until they eat all the plants.

4. People can't live on the land because cattle

5. People blow up a hill to build a road through it.

Effect

1. _____

2. A high wind removes topsoil, lowering the land.

3. _____

4. _____

5. _____

Our Earth's Resources

The planet we live on is not only beautiful, but it also has many resources. These resources make life on Earth possible and pleasant. Use the clues below and the words in the box to learn the name of each resource.

water
heat
light
plants
soil
animals
oxygen
carbon dioxide

1. Plants and animals need this liquid to live. _____

2. Some members of this group of living things are so small that you cannot see them easily. Some are among the largest living things on Earth. Most have roots. _____

3. This huge group includes tiny insects and large mammals. _____

4. We breathe this gas; without it, animals would die. _____

5. Plants take in this gas; without it, they would die. _____

6. Plants grow in this; it has many nutrients they need. _____

7. This comes from the sun, and plants need it to make their own food. _____

8. This comes from the sun and keeps our planet from being too cold. _____

Ecology

Ecology is the study of the connection among all living things and their environment. All living things in a habitat depend on each other for food. The link is more complicated than that, though. All living things on Earth depend on each other. If any one part of the web of life takes more than it should, others will die.

Learn more about ecology by choosing a sentence from the box to complete each cause and effect statement.

Cause

1. If humans drain water from a lake faster than the water cycle can replace it,
2. If humans cut down rain forests to graze cattle,
3. If there is too little carbon dioxide in the air,
4. Rain forest animals will die
5. We study ecology because

Effect

1. _____
2. _____
3. _____
4. _____
5. _____

---

we need to learn how all living things depend on each other.
animals that breathe oxygen will die.
there will be fewer plants to exchange carbon dioxide for oxygen.
the lake may dry up, and animals that live in or near it may die.
if the trees in their habitat are cut down.

What Can Happen?

Earth has a layer of gases around it called the atmosphere. Heat and light can come through the atmosphere, but it keeps helpful gases inside and harmful ones outside. These gases help keep Earth's temperature at a level that can maintain life. Today, the amount of gases in the atmosphere is increasing, causing temperatures on Earth to rise. This is called global warming.

To learn more about the effects of increased temperatures on Earth, write the number of the picture in front of the sentence that describes it.

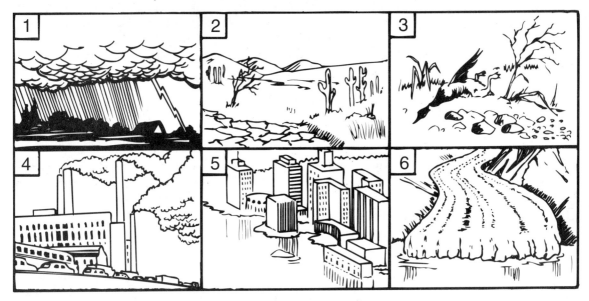

_____ In some areas, less rain will fall, causing droughts.

_____ Polar ice begins to melt.

_____ Factories and automobiles produce carbon dioxide.

_____ The rise in Earth's temperature will cause big storms.

_____ Ocean water levels rise and flood lowland areas, including cities and islands.

_____ Because of the change in the temperature, many plants and animals will die.

Pull-Out Answers

Page 1
A. 4, 3, 2, 1, 5
B. 5, 2, 4, 6, 3, 1

Page 2
1. a mosquito flies out.
2. The babies learn to fly, and then they leave home.
3. Answers will vary but must show logical progression between puppy stage and adulthood for the dog. All information should be expressed as complete sentences.

Page 3
1. The white rabbit sat still on the snow.
2. a predator would leave it alone.
3. the frog caught an insect that hadn't seen the frog.
4. the bird did not eat it.

Page 4
1. bee
2. rattlesnake
3. hermit crab
4. puffer
5. turtle
6. skunk
7. horned lizard

Page 5

Page 6
1. The child has her mother's eyes.
2. The child has her father's hair color.
3. The child has her mother's nose shape.
4. The child has her mother's freckles.
5. The child has her father's mouth shape.
6. The parents' genes determined that their offspring would be a human.
7. Answers will vary.

Page 7
1. mom
2. dad
3.–8. Answers will vary.

Page 8
2, 1, 4, 3; Answers will vary but should include any of the facts listed on the page.

Page 9
1. desert
2. arctic
3. pond
4. under a rock
5. temperate forest

Page 10
Answers will vary.

Page 11
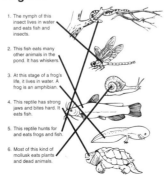

Page 12
Answers will vary somewhat but should generally follow this guide:
1. The sun's energy in the form of light hits Earth.
2. Plants use this light energy to grow and live.
3. Herbivores eat plants for energy.
4. Carnivores eat herbivores for energy.
5. Omnivores eat herbivores and carnivores.

Page 13
1. skunk, omnivore
2. snail, herbivore
3. bear, carnivore
4. koala, herbivore
5. mockingbird, omnivore

Page 14
1. All animals should be listed.
2. A blue circle should be drawn around all living things in the picture.
3. A red line should be drawn under all animals.
4. A yellow X should be drawn on all plants and animals.
5. Answers will vary somewhat but should mention the oxygen the plants give off.

Page 15

Pull-Out Answers

Page 16
a. squirrels
b. mice, rabbits
c. worms, robins
d. deer
2. foxes, hawks, owls
3. Answers will vary.

Page 17
1. It dies.
2. It dies.
3. They die.
4. It dies.
5. Answers will vary but should have to do with the interdependence of plants and animals and the need for clean air and water in an environment.

Page 18
1. Plants need clean air to flourish; Circle: Healthy plants grow nicely in polluted air.
2. Circle: All animals like to eat other animals; Some animals eat plants, and others eat animals.
3. A large habitat is called a biome; Circle: A wet biome is the best place for animals to live in.
4. If plants in a habitat die, eventually, both plant- and animal-eating animals die; Circle: If plants in a habitat die, only plant-eating animals will die. The other animals will live.
5. Circle: If the air is polluted, only plants will get sick; Polluted air affects both plants and animals.
6. Circle: People are not affected by polluted water; Polluted water is unhealthful for all animals and plants.

Page 19
Answers will vary.

Page 20

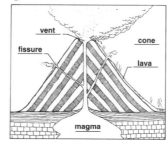

Page 21
1. The heavy soil in the cliff slides down on the road below.
2. When the tsunami hits land, it washes away houses, other buildings, cars, and trees.
3. Part of the highway has fallen down.

Page 22
Answers will vary, but suggestions follow:
1. Both destroy the land and change the way it looks.
2. Water and wind take away soil, making it difficult for plants to grow.
3. When the land cannot absorb any more water, the water runs off the land, taking a layer of soil with it.
4. As a glacier moves over land, the rocks in the ice grind away solid rock and carve deep valleys.

Page 23
1. Heavy jungle rains wash away the topsoil.
2. Someone scrapes all grasses and plants from the land.
3. The land becomes a desert.
4. turned the land into a desert.
5. Wind hollows out the hill on either side of the road.

Page 24
1. water   2. plants
3. animals   4. oxygen
5. carbon dioxide
6. soil
7. light
8. heat

Page 25
1. the lake may dry up, and animals that live in or near it may die.
2. animals that breathe oxygen will die.
3. there will be fewer plants to exchange carbon dioxide for oxygen.
4. if the trees in their habitat are cut down.
5. we need to learn how all living things depend on each other.

Page 26
1. The rise in Earth's temperature will cause big storms.
2. In some areas, less rain will fall, causing droughts.
3. Because of the change in the temperature, many plants and animals will die.
4. Factories and automobiles produce carbon dioxide.
5. Ocean water levels rise and flood lowland areas, including cities and islands.
6. Polar ice begins to melt.

Pull-Out Answers

Page 27

1. The temperature of the ocean rises near the equator. High winds begin to turn in a circle. At the center of this storm system is an eye, or calm place. This storm gathers strength as it moves over warm water. It produces heavy rain, high waves, and great wind. It is called a typhoon or a hurricane.
2. This storm forms when cold air and hot air meet. The air temperature changes quickly and in the northern part of North America, heavy snow falls.
3. This storm forms when air rises rapidly and cool air plunges to the ground. The air temperature changes quickly, and in the southern part of North America, thunderstorms occur and may include hail or rain.
4. This type of storm occurs over land. It begins when warm, humid air rises rapidly and more warm air rushes in to replace it. When the thunderstorm moves in a slow circle high above Earth, tornadoes can drop to Earth. This twisting storm moves across the land with high, twisting winds.

Page 28

1. Help the owners of the factory learn how to dispose of toxic chemicals properly.
2. We should encourage automobile manufacturers to make cars that use fuels that will not add carbon dioxide to the atmosphere.
3. Manufacturers began making air conditioners and refrigerators that use a safe chemical.

Page 29

metal: soda cans, food cans, aluminum foil; paper: paper towels, newspapers, magazines, napkins; plastic: plastic soda bottles, plastic rings for soda cans; glass: food jars, broken drinking glasses

Page 30

1. The boy moves his hand away and cries.
2. A message goes from Ellie's nose to her brain that tells Ellie that her favorite dinner is cooking.
3. Andy brings the hamster to the teacher.

Page 31

1. 2, 1, 3
2. 3, 1, 2
3. 1, 3, 2
4. 3, 2, 1

Page 32

Answers will vary.

Page 33

Answers will vary, but suggestions are as follows:

1. Sound waves hit the outer ear and move into the ear canal.
2. Sound waves vibrate the eardrum.
3. Tiny hairs send messages to the brain about the sound waves.

Page 34

Answers will vary.

Page 35

Use the classifications that follow as guidelines only. A child may reasonably categorize things in different ways. If it's logical, allow it. hot: dog's tongue, boiling water, fire; wet: dog's tongue, cat's tongue, water, milk, snow, rain, ice, boiling water, puppy's teeth; rough: cat's tongue, sandpaper, tree bark; soft: cat's fur, whiskers, rose petal, balloon, feather, dog's tongue; cold: ice, snow; sharp: puppy's teeth, tip of a nail, thorn, fork tines

Page 36

bucket, big rock, small rock, bike helmet, bike, skateboard, tree, grass

Page 37

Page 38

1. cloud
2. oxygen
3. carbon dioxide
4. carbon dioxide
5. oxygen
6. steam
7. air

```
n c m i k w r l c g s o m
m t n e g y x o l h u s a
c a r b o n d i o x i d e
k l e p d x n b u y o j t
e a i r i k a e d r a e s
```

Page 39

solid: ice, wood, pen; liquid: water, milk, coffee, tea; gas: oxygen, steam, carbon dioxide

Page 40

1. An ice cube is a solid. Heat from the sun changes the solid form of water to liquid (water). The sun's heat changes the liquid to a gas.
2. Water is in the form of a gas in the cloud. Rain (liquid) falls from the cloud into a lake. Then the cold temperature makes the lake turn to ice.
3. Answers will vary but should include information about heat/cold changing the state of matter.

Pull-Out Answers

Page 41
Answers will vary, but the child should have collected five different rocks and should have given each the scratch test using the suggested tools. A glass pocket mirror can be used in place of a scratch plate.

Page 42
Answers will vary but should be correctly based on the scratch scores from page 41. The line graph should correctly reflect the information.

Page 43
1. The poles will repel each other. The child's drawing should reflect that.

2. The two magnets will be attracted to each other.

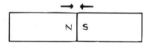

3. The prediction, based on the first question, should say that the poles will repel each other.
4. The poles repelled each other.
5. Answers will vary.

Page 44
The magnet picked up only the objects containing steel or iron.

Page 45
1.–2. Depending on the strength of each magnet, the magnets should still attract each other, north pole to south pole.
3.–4. The like poles should still repel each other, depending on the strength of the magnets.
5. The magnet will not attract the toothpick in the foil because neither has iron or steel.
6. The magnet will not attract the wood because a magnet is not attracted to wood.
7. The magnet is not attracted to the foil because it contains neither iron nor steel.

Page 46
1. Answers will vary depending on the strength of the magnet.
2. Answers will vary depending on the strength of the magnet, but it should not affect the magnet's field.
3. Answers will vary depending on the strength of the magnet.
4. Answers will vary depending on the strength of the magnet.
5. It helped them to arrive at the place they wanted to go by showing them directions.

Page 47

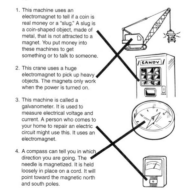

1. This machine uses an electromagnet to tell if a coin is real money or a "slug." A slug is a coin-shaped object, made of metal, that is not attracted to a magnet. You put money into these machines to get something or to talk to someone.
2. This crane uses a huge electromagnet to pick up heavy objects. The magnets only work when the power is turned on.
3. This machine is called a galvanometer. It is used to measure electrical voltage and current. A person who comes to your home to repair an electric circuit might use this. It uses an electromagnet.
4. A compass can tell you in which direction you are going. The needle is magnetized. It is held loosely in place on a cord. It will point toward the magnetic north and south poles.

Page 48
1.–5. Answers will vary.
6. The bike moves a person.

Page 49
Inclined planes include the ladder, the ramp by the loading dock, the ramp leading up to the porch, the board leaning on the curb, and the roof.

Page 50
Answers will vary.

Page 51
The child should draw the rest of the picture to show an understanding of how the pulley works.

Page 52
1. wheel and axle
2. lever
3. inclined plane
4. lever
5. lever
6. wheel and axle
7. pulley
8. wheel and axle
9. wheel and axle
10. lever

Bad Weather

Every year all over Earth, bad weather occurs. All weather is caused by the sun's heat. A rise of only a few degrees can cause many problems, including the bad weather described below. Draw a line from each picture of severe weather to the description of the weather and its cause.

1. The temperature of the ocean rises near the equator. High winds begin to turn in a circle. At the center of this storm system is an eye, or calm place. This storm gathers strength as it moves over warm water. It produces heavy rain, high waves, and great wind. It is called a typhoon or a hurricane.

2. This storm forms when cold air and hot air meet. The air temperature changes quickly and in the northern part of North America, heavy snow falls.

3. This storm forms when air rises rapidly and cool air plunges to the ground. The air temperature changes quickly, and in the southern part of North America, thunderstorms occur and may include hail or rain.

4. This type of storm occurs over land. It begins when warm, humid air rises rapidly and more warm air rushes in to replace it. When the thunderstorm moves in a slow circle high above Earth, tornadoes can drop to Earth. This twisting storm moves across the land with high, twisting winds.

Taking Care of Our Environment

Because all plants and animals depend on each other, we should take care of our environment to help keep plants and other animals alive. Doing this will also keep us alive. Read each paragraph and choose the sentence that tells how to help the environment.

1. A factory dumps toxic chemicals it has used into a nearby river. Some fish die. Others get sick. Animals that eat fish from this river get sick or die.

    ○ Close down the factory, putting a lot of people out of work.

    ○ Help the owners of the factory learn how to dispose of toxic chemicals properly.

2. Automobile exhaust adds carbon dioxide to our atmosphere. Adding this gas to the atmosphere causes global warming. Some parts of the world are so dry that plants die.

    ○ We should make it illegal to drive any automobile.

    ○ We should encourage automobile manufacturers to make cars that use fuels that will not add carbon dioxide to the atmosphere.

3. Refrigerators and air conditioners once used a chemical that damaged the layer of gases around Earth. It destroyed a part of the protective gases and let in harmful rays from the sun. What did we do?

    ○ We banned all refrigerators and air conditioning.

    ○ Manufacturers began making air conditioners and refrigerators that use a safe chemical.

What Can You Do?

Recycling helps the environment. You can help, too. You can recycle. Recycling means collecting used paper, metal, glass, and plastic in special containers so that it can be made into other items. This helps preserve some of Earth's resources. Collect used containers or paper around your home for recycling. To practice, write the name of each product in the box on the correct container.

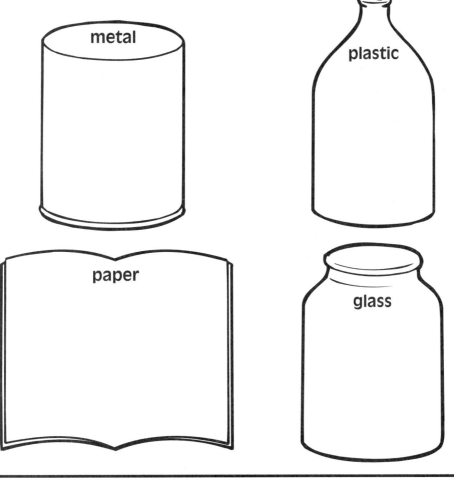

food cans	soda cans	plastic soda bottles
food jars	newspapers	broken drinking glasses
magazines	paper towels	plastic rings for soda cans
napkins	aluminum foil	

Your Brain

Your brain receives electric messages from nerves in all parts of your body. It then sends out messages to tell those body parts what to do.

Read the paragraphs below. Fill in the circle that best predicts what will happen next.

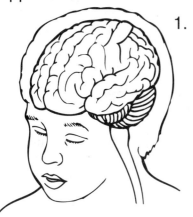

1. A baby boy touches a hot burner on a stove. The nerves in his fingers zip a message to his brain: "That's hot. We're injured." The boy's brain shoots a message to his hand that says, "Move away from the heat."

   ○ The boy moves his other hand to the hot stove and laughs.

   ○ The boy moves his hand away and cries.

2. Mother is cooking Ellie's favorite dinner. Good smells float through the house. Ellie smells the dinner.

   ○ A message goes from Ellie's nose to her brain that tells Ellie that her favorite dinner is cooking.

   ○ A message goes from Ellie's nose to her brain to leave the house to escape from danger.

3. Andy's teacher asks him to bring the class hamster to the reading group. The sounds go into Andy's ears, and nerves send an electronic message to his brain.

   ○ Andy covers the birdcage.

   ○ Andy brings the hamster to the teacher.

Seeing Is Believing

Your eyes allow you to use one of your five senses—sight. When you see something, beams of light come in through the pupil. The lens and the cornea focus the image. At the back of your eyeball is your retina. The optic nerve connects the retina to the brain. It translates images seen by the eye into electric messages and then sends them to your brain. Your brain then tells other parts of your body what to do by sending more electronic messages.

Number the sentences in each set in 1, 2, 3 order.

1. ____ A message goes from your eyes to your brain, "Your friend is in trouble."
   ____ You see your friend fighting with someone.
   ____ Your brain sends a message to your muscles, and you walk to your friend.

2. ____ Your brain sends a message to your muscles, "Get away from this animal."
   ____ You see a poisonous snake asleep on a rock.
   ____ Your retina sends an electronic message to your brain.

3. ____ You are reading a book when the sun goes down.
   ____ You turn on the electric light so you can see the book.
   ____ Your eyes send a message to your brain, "It's too dark to see this book."

4. ____ Messages from your brain to your hands tell you where to put them to catch the ball.
   ____ A message from your eyes to your brain says, "The ball is up in the air."
   ____ You throw a ball in the air.

And Then What?

When you breathe in, odors stick to some tiny hairs at the base of your nose. These send electronic messages to your brain to tell it about the odor. Your brain tells other parts of your body what to do.

Read each sentence below about odors. Write a sentence telling what your brain would tell you to do.

1. You smell raw onions. _____

2. The house catches fire while you are asleep. You smell smoke.

3. Your little brother ate a chocolate cookie in the kitchen. You smell the chocolate when he comes near you. _____

4. Your dad put on shaving cologne. He gives you a hug, and you smell it. _____

5. Your mom brings home a pizza. When she opens the box, you smell it. _____

6. In your own words, tell how your nose and your brain work together to help you smell. _____

Let's Hear It for the Ears!

Your ears have a special shape that funnels sound. When sound waves hit the outer edge of your ear, they move into the ear canal. Then they strike the eardrum, which vibrates and sends the vibrations to the inner ear. From there, tiny hairs send electronic messages to the brain about the sound waves. The brain sends messages to other parts of the body about the sound.

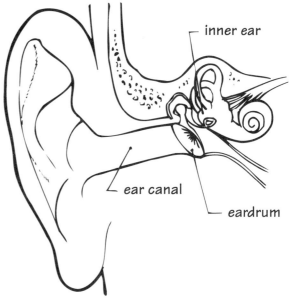

Use the words in each drum to write a complete sentence about hearing.

1. ear / ear canal

2. eardrum / vibrate

3. sound waves / messages

Mmmmm Good!

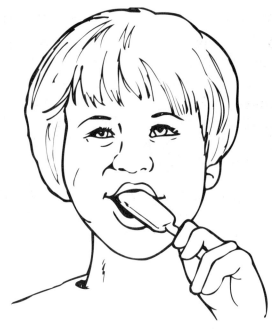

Taste is one of your five senses. On your tongue, you have taste buds. These are grouped by areas. On one part of your tongue, you taste salty things. On other parts of your tongue, you taste sweet, sour, and bitter. People's taste buds differ from person to person. Therefore, people don't taste things exactly the same way. Maybe you like onions, but your friend does not. Your taste buds send messages to your brain about tastes. Your brain sends messages to your body about tastes.

1. Name a food you don't like. _____
2. Is this food sweet, sour, salty, or bitter? _____
3. What do your taste buds do when you put that food in your mouth?
   _____
4. What would your brain tell you to do? _____
5. Name one salty food you like to taste. _____
6. Name one sweet food you like to taste. _____
7. Name a sour food you don't like to taste. _____
8. Name a bitter food you do not like to taste. _____

In your own words, tell what happens when you put a lemon in your mouth. Include these words: **taste buds**, **electronic messages**, **brain**.

_____
_____
_____

It's Touching

Your sense of touch lets you feel the texture of things. With this sense, you can feel heat, cold, pain, and pressure. When you touch something, nerves in the layer of skin called the dermis send messages to your brain. Some parts of your body have more nerves than others. That makes them more sensitive than parts that have fewer nerves.

Write each word from the box in the correct place to show how it would feel if you touched it.

feather	cat's fur	rose petal	balloon	puppy's teeth
thorn	fork tines	tip of a nail	sandpaper	cat's tongue
tree bark	water	milk	ice	dog's tongue
rain	fire	snow	whiskers	boiling water

© Frank Schaffer Publications, Inc.   35   FS111161 Science

What's the Matter?

Everything is made of matter. Matter is anything that takes up space and has mass. Mass is the amount of matter in an object. Matter is made of tiny particles. Matter has three forms: solid, liquid, and gas. A solid has its own shape. Particles in a solid vibrate in place. They are so tiny that you cannot see them. If you move a solid, it will keep its shape. You can turn it upside down, and it will still keep its shape. Circle all of the solids in the picture below.

It Pours

Liquid is a form of matter. It is made of particles that slide slowly from side to side, over and under each other. Liquid does not have a shape of its own. It takes the shape of whatever it's in. If you put liquid in a cylinder, it takes that shape. If you pour the liquid from the cylinder to a rectangular container, it will take that shape. If you turn the container on its side, the liquid will pour out.

Follow the names of liquids in the maze to find your way to the end.

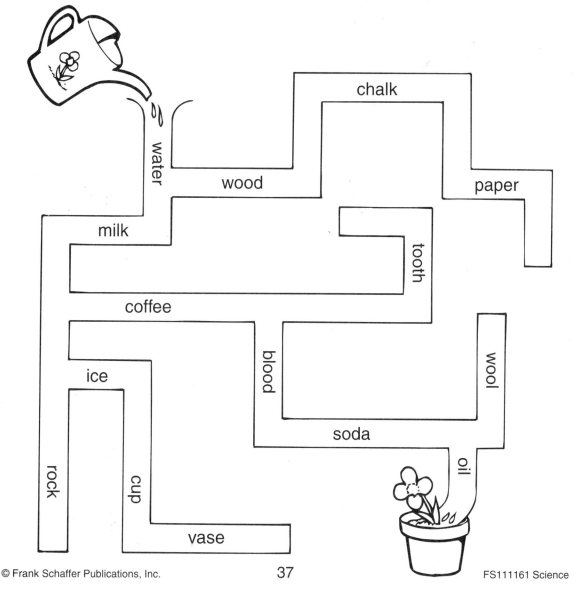

Looking for Gas

Gas is a form of matter. It is made of tiny particles that move fast. A gas has no shape or size. Heat can change a solid to a liquid and a liquid to a gas. Use the words in the box to name each type of gas. Some words will be used more than once.

1. You can see this form of gas in the sky. Water droplets can gather here and produce rain. _____

2. Animals breathe this gas. _____

3. Plants breathe this gas. _____

4. Animals exhale this gas as waste. _____

5. Plants give off this gas. _____

6. This gas comes off the top of boiling water. _____

7. This gas surrounds us. _____

| carbon dioxide | oxygen | cloud | air | steam |

Find and circle the words from the box in the puzzle.

n	c	m	i	k	w	r	l	c	g	s	o	m
m	t	n	e	g	y	x	o	l	h	u	s	a
c	a	r	b	o	n	d	i	o	x	i	d	e
k	l	e	p	d	x	n	b	u	y	o	j	t
e	a	i	r	i	k	a	e	d	r	a	e	s

Comparing Matter

Matter comes in three forms: solid, liquid, and gas. Using the words in the box, write the names of the solids on the trash can; write the names of the gases on the cloud of steam; write the names of the liquids on the drop of water.

Write one more in each group.

| ice | steam | water | wood | milk |
| coffee | oxygen | tea | pen | carbon dioxide |

Temperature Matters

Look at each picture to see one way that temperature changes matter. Write three complete sentences for each picture to describe what is happening.

1.

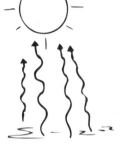

_____
_____
_____

2.

_____
_____
_____

3. Explain how temperature affects the state of matter.

_____
_____
_____

Hard to Figure

You know that a rock is a solid, but there are differences among rocks. For example, some rocks are harder than others. One way to compare the hardness of rocks is by using the Mohs Hardness Scale. Using this method, you scratch each rock with several tools. The results help you determine the hardness of the rocks.

To discover the hardness of some rocks, collect five different kinds of rocks. Ask an adult to help you. Try to scratch each rock with these different tools: 1. fingernail, 2. a penny, 3. a streak plate (if available, this is a piece of glass used for this purpose), and 4. a steel nail. (Always use these tools in the order listed.) The softest rocks can be scratched with your fingernail. The hardest rocks will not be scratched with any of these tools.

Rock	Did it scratch? (yes or no)	What scratched it?	Description of rock
1.			
2.			
3.			
4.			
5.			

Taking Another Look

Use your own hardness scale to compare the rocks you tested on page 41. Give rocks you scratched with your fingernail a score of 2.5; rocks that you scratched with a penny a 3; those you could scratch with the streak plate a 5.5; and those you scratched with a steel nail a 6.5. Give those rocks you could not scratch with any of those tools "greater than 6.5." Plot the score of each rock on the line graph to show how your rocks compare.

Hardness

greater than 6.5					
6.5 or less					
5.5 or less					
3.00 or less					
2.5 or less					
Rocks:	1	2	3	4	5

Magnetic Poles

To complete this page, you will need two magnets. Every magnet has a north pole and a south pole. These are marked as N and S. Use your magnets to follow the instructions below.

1. Put your magnets together, north pole next to north pole. Write what happened in your own words in complete sentences. Draw a picture of what happened. Be sure to label the north pole with an N and the south pole with an S.

   _____
   _____
   _____

2. Put your magnets together, south pole next to north pole. Tell what happened using complete sentences. Draw a picture of what happened. Be sure to label the north pole with an N and the south pole with an S.

   _____
   _____
   _____

3. What do you think would happen if you put south pole next to south pole? _____
   _____

   What made you come to that conclusion? _____
   _____

4. Now put the two magnets together, south pole to south pole. What happened? _____
   _____

5. Did you predict accurately or not? _____

That Certain Attraction

A magnet will pick up an object made of steel or iron. Collect the following items: a magnet, a paper clip, a button, a nail, a toothpick, a sponge, a paper napkin, staples, a fork or spoon, scissors, a penny, a quarter, a rubber band, and a book. Put items that you think the magnet can pick up in one pile. Put the other objects in a second pile. Write the names of all objects below and record whether you think the objects can be picked up by a magnet. Try to pick up each object using the magnet. Write what happened.

Name of the object	Prediction	Result
1.		
2.		
3.		
4.		
5.		
6.		
7.		
8.		
9.		
10.		
11.		
12.		
13.		

Explain why you could pick up some things with a magnet but could not pick up others. _____

How did you do with your predictions? _____

Using the Force

For this page, you will need two magnets, a tissue or a paper towel, a toothpick, and a piece of aluminum foil.

1. Wrap each magnet in a piece of tissue or paper towel. Mark an S and an N on the paper to tell where the north and south poles are. Move the magnets toward each other, north pole to south pole on each side. What happened? _____

2. How much did the paper affect the magnetic attraction of the north and south poles? _____

3. With the magnets still wrapped in paper, push the south pole of one toward the south pole of the other. What happened? _____

4. Remove the paper from the magnets. Did the magnets act the same as they did without the paper between them? _____

5. Wrap a toothpick in aluminum foil. Will a magnet attract the toothpick? _____ Why? _____

6. Remove the aluminum foil. Hold the magnet near the toothpick. Is the magnet attracted to it? _____ Why? _____

7. Is the magnet attracted to the aluminum foil? _____ Why or why not? _____

Make Your Own Compass

A compass is a tool we can use to find directions. In a compass, a magnetized needle floats in a liquid or is attached at its center so it can turn easily. A compass often has a circular shape, and it is marked on its outside with directional symbols: N, S, W, E. For this lesson, you will need a steel needle, a bar magnet, glue, a pencil, thread, and a drinking tumbler made of glass or plastic.

Put one dot of glue on the middle of the needle. Put one end of a 4-inch length of thread on the glue. Be sure the glue does not touch anything else. (You can do this by sticking the needle in a cork before you begin.) Let this dry overnight. Tie the other end of the thread to the middle of a pencil. The thread should be only long enough to suspend the needle in the glass without touching the sides. Stroke one end of the needle on the bar magnet about 30 times in one direction only. This will magnetize your needle. Suspend the needle in the glass.

1. Move the magnet a foot away from your needle. What happened? _____

2. Put a piece of paper between your compass and the magnet. What happened? _____

3. Put a small book between your compass and the magnet. What happened? _____

4. Put a thick book between your compass and your magnet. What happened? _____

5. Sailing ship crews in the 12th century probably used the first compass. How do you think it would have helped them? _____

Meet Some Real Magnets

Magnets play a big role in our everyday lives. Draw a line to match each picture to a description of the way a magnet is used to make it work. Some machines use electromagnets. These magnets work only when electricity powers them.

1. This machine uses an electromagnet to tell if a coin is real money or a "slug." A slug is a coin-shaped object, made of metal, that is not attracted to a magnet. You put money into these machines to get something or to talk to someone.

2. This crane uses a huge electromagnet to pick up heavy objects. The magnets only work when the power is turned on.

3. This machine is called a galvanometer. It is used to measure electrical voltage and current. A person who comes to your home to repair an electric circuit might use this. It uses an electromagnet.

4. A compass can tell you in which direction you are going. The needle is magnetized. It is held loosely in place on a cord. It will point toward the magnetic north and south poles.

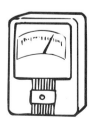

Simple Machines: Wheel and Axle

Machines make everyday life easier than it would be without them. One of the four kinds of simple machines is the wheel and axle. This simple machine allows you to move something with only a fraction of the effort you might have made without it. For example, if you try to push a huge box full of heavy rocks, you would have a tough time. Friction prevents the box from moving smoothly. Tiny irregularities on the surface of the ground and the surface of the box catch each other, and hinder the box from easily moving. If you put the box on wheels, you can push or pull it much easier.

List five things you and your family use that have wheels.

1. _____
2. _____
3. _____
4. _____
5. _____
6. A bike has wheels. What does a bike move? (Use a complete sentence.) _____

© Frank Schaffer Publications, Inc. 48 FS111161 Science

Simple Machines: The Inclined Plane

A plane is a surface, like the ground. An inclined plane is a ramp. A ramp allows you to move things from one level to a higher level with less effort than you would use if you had to lift it or push it. Look at the picture below. Circle all of the inclined planes.

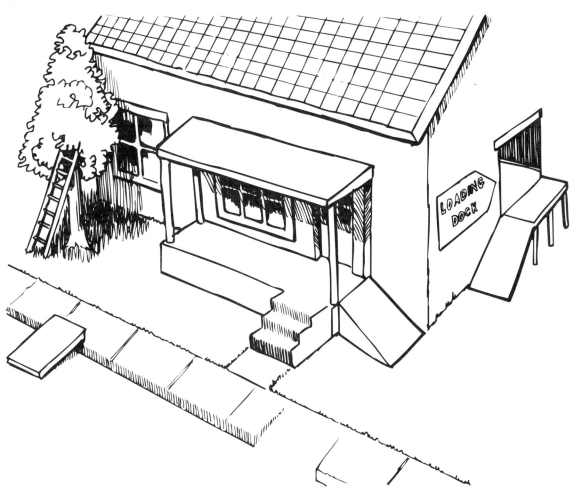

Choose one of the inclined planes in the picture and explain in your own words how the ramp helps someone use less energy.

_____

_____

Simple Machines: The Lever

A lever has two parts: a rigid bar and a fulcrum. The fulcrum does not move, but the bar does. Look at the tools below. Choose three of them and write how they make it easier for people to work or play. Use complete sentences.

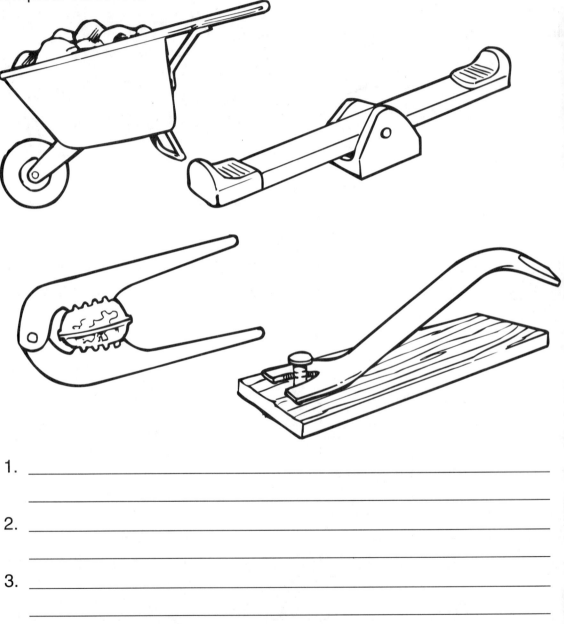

1. _____
   _____
2. _____
   _____
3. _____
   _____

© Frank Schaffer Publications, Inc. FS111161 Science

Simple Machines: The Pulley

A pulley is a simple machine. It helps someone pull down to lift up. The pulley below is labeled to show you the parts. Using a pulley, you can climb to the top of a ladder and lift a paint can. To do this, you simply pull down on a rope that runs over the top of a wheel. Using pulleys, people who wash the windows of tall buildings can pull themselves up with only a little effort.

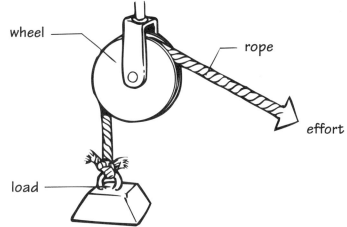

Below is a picture of a heavy box that you want to lift to an apartment on the fifth floor of a building. Draw the details on the pulley. Remember to use a wheel and rope.

It's Simple

Below are pictures of simple machines. On the line next to each picture, write the type of simple machine it is. Use these words: **inclined plane**, **pulley**, **lever**, **wheel and axle**.

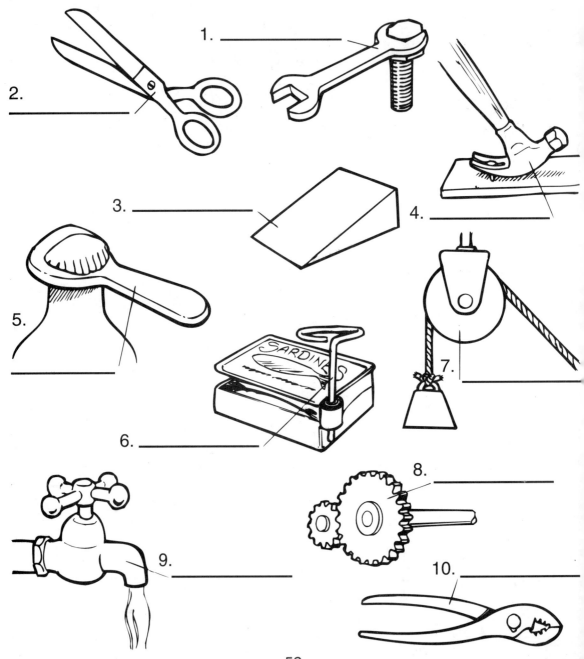